SCIENCE INVESTIGATORS

Investigating LIGHT and SOUND Through Modeling

Kristin Thiel

Cavendish Square
New York

Published in 2020 by Cavendish Square Publishing, LLC
243 5th Avenue, Suite 136, New York, NY 10016

First Edition

Cataloging-in-Publication Data
Names: Thiel, Kristin.
Title: Investigating light and sound through modeling / Kristin Thiel.
Description: New York : Cavendish Square, 2020. | Series: Science investigators | Includes glossary and index.
Identifiers: ISBN 9781502652461 (pbk.) | ISBN 9781052652478 (library bound) | ISBN 9781502652485 (ebook)
Subjects: LCSH: Light--Juvenile literature. | Sound--Juvenile literature.
Classification: LCC QC360.T44 2020 | DDC 535--dc23

Editor: Caitlyn Miller
Copy Editor: Nathan Heidelberger
Associate Art Director: Alan Sliwinski
Designer: Ginny Kemmerer
Production Coordinator: Karol Szymczuk
Photo Research: J8 Media

The photographs in this book are used by permission and through the courtesy of:
Cover, redsnapper/Alamy Stock Photo; p. 4 Elizabeth Foster/Shutterstock.com; p. 8 Marc Romanelli/Getty Images; p. 10; Robin Lund/Shutterstock.com; p. 12 Jena Ardell/Moment/Getty Images; p. 14 (top) Oren Jack Turner, Princeton, NJ/Library of Congress/File: Albert Einstein 1947.jpg/Wikimedia Commons/Public Domain; p. 14 (bottom) Colin, Own work/File: Josselin Château Evening Light Reflected 2016-08-15 WLM.jpg/Wikimedia Commons/CC by 3.0; p. 15 Pat_Hastings/Shutterstock.com; p. 16 gritsalak karalak/Shutterstock.com; p. 19 Diana Haronis dianasphotoart.com/Moment/Getty Images; p. 22 Rebecca Drobis/Getty Images; p. 25 pablofdezr/Shutterstock.com; p. 26 Designua/Shutterstock.com; p. 28 Angyalosi Beata/Shutterstock.com; p. 30 Fouad A. Saad/Shutterstock.com; p. 32 Stocktrek Images/Getty Images; p. 34 Eric Meola/The Image Bank/Getty Images; p. 36 MIKKEL JUUL JENSEN/Science Source; p. 37 zaferkizilkaya/Shutterstock.com; p. 39 OLIVER-stockphoto/Shutterstock.com; p. 41 Evannovostro/Shutterstock.com.

Printed in the United States of America

Contents

We see colors thanks to light waves.

CHAPTER 1

Key Concepts

Most of us only think about light and sound in big, general ways. If a room is dark, we turn on the light. If we cannot hear the television, we turn up the volume. However, light and sound work in many other ways. For example, light waves and sound waves are busy all around you even as you sit quietly reading.

You are able to read this book because of light. It's not just because you are sitting by a window or near a lamp. Light is reflecting off the white paper. Light is also being absorbed by the black letters. That contrast allows you to see each letter and each word.

Your teacher may ask to "read silently." However, there is nothing silent about reading. When your fingertips touch the page, there is a tiny sound. When you turn the page, the paper rustles. You know all this because sound travels through the air to your ears.

COMPARING LIGHT AND SOUND

Light and sound both travel as waves. We can imagine them looking like waves in the ocean. They move forward. They stretch up, and they plunge down. However, light and sound waves are also very different from each other.

Light waves are made of photons. Photons are particles of energy. This means light behaves like a wave and like a particle. Sound waves, on the other hand, are caused by vibrations. Vibrations happen when a particle moves up and down or side to side.

Light travels a lot faster than sound. You will see something before you hear it. You will always see lightning before you hear the slower-moving sound of thunder. Sound, however, is not always stopped by objects in its way. Light is often stopped by objects. You can shut your bedroom door and not see your family. You will still be able to hear them.

AMAZING WAVES

Sound and light waves are part of everyday life. They also are extraordinary! There are many examples of sound and light waves making a big difference in our world. Sometimes they even save lives. Sound waves allow people to call to each other for help. Light waves can be used in medicine.

The Iceberg and the Radio

Sound saved lives when the *Titanic* sank. The *Titanic* was a ship that was supposed to be unsinkable. On its first trip, it hit an iceberg—and it started sinking. More than 1,500 people died. However, more than 700 survived. Many credit the power of sound with helping those people.

The *Titanic* sank in 1912, when wireless communication was new. Before, people communicated between ships by shouting. As you can imagine, that was not very effective. Two ships would have to be very close together for the sound waves of a shout to reach another person. (Sometimes when you try to say hi to your friends at school, they cannot hear you—even when you are in the same hallway.) The *Titanic* had a new communication system on board, which converted radio waves into sound waves. The crew of the *Titanic* could send emergency calls to a ship 500 miles

Understanding Sound with a Violin

Models help scientists understand and explain our world. Models can be computer simulations, creations made from everyday objects, mathematical equations, or charts and graphs. There are lots of ways to model sound waves. Some important early mathematicians proved that. Daniel Bernoulli and Jean Le Rond d'Alembert were two such mathematicians.

Understanding how instruments work can help us understand the science of sound.

Bernoulli was a mathematician who lived in Switzerland in the early 1700s. He learned about sound by studying violins. He believed that understanding how violin strings made sound would help him understand how other things made sound too.

He imagined that each string was made up of different segments. This helped Bernoulli think about each part of the string as having mass. "Mass" means how much matter something has. Everything that takes up space has mass that can be measured.

Everything that has mass follows something called the laws of motion. Sir Isaac Newton, a physicist who studied matter and how things interact with each other, was the first person to describe these rules. Bernoulli used Newton's laws of motion to calculate the shape of the vibrations of violin strings. Bernoulli could then figure out how that looked as sound waves traveling to ears.

Many years later, Jean Le Rond d'Alembert, a mathematician in France, developed Bernoulli's ideas. He figured out that tension—how tight the string was pulled—affected the sound waves the string produced. In other words, he began to understand the math and science behind making different notes.

(804 kilometers) away. (This is a lot farther than the distance between you and your friends in the school hallway!)

The Power of Light

Sound waves are not the only life-changing waves out there. Scientists have learned how to control light waves to do amazing things. Today, lasers are changing lives in many ways. The word "laser" is an acronym. Each of its letters stands for a word. Put together, these words define what a laser is. A laser is Light Amplification by Stimulated Emission of Radiation.

Lasers are incredibly powerful and have many uses.

Lasers are light, but they do not occur in nature. Lasers are a human invention, and laser light is known for being very narrow and bright. We can focus lasers so the light hits only one small area.

Laser light is a strong, precise tool. Doctors use it to cut through human tissue during surgery. Jewelers use it to cut through diamonds. Lasers also help us measure the distance between things that are very far apart. For example, NASA has measured the distance between Earth and the moon with laser light. NASA also uses lasers to "see" inside rocks on other planets. Lasers even carry and record information. For example, laser light can carry internet signals.

The sun provides Earth with the light and heat that makes life possible on our planet.

CHAPTER 2

Modeling Properties of Light Waves

Light is a basic part of everyday life, like air and food. The sun shines every day, even on cloudy days. At night, electric light or fires give us light. Light is common, but it is also complex.

THE DEBATE ABOUT WHAT LIGHT IS

Light does not have mass. We think of everything that exists as being made up of matter. Light is an exception to this. It exists, but it is not made like anything else we know. It also moves like nothing else. When nothing slows it down, the speed of light is 186,282 miles per second (299,792 kilometers per second).

That is about a million times faster than sound. It takes a mere eight minutes for this incredible energy to travel from the sun to Earth.

Defining what light is has caused big arguments. In the 1600s, scientists fought about light being a particle or a wave. In the early 1900s, even Albert Einstein got involved. Both of the definitions of light are correct. Neither of these ideas, by itself, tells the whole story. We need to consider light as both a particle and a wave to understand it.

Albert Einstein studied physics and was interested in the properties of light.

The way light interacts with water causes reflections.

Light as a Wave

Thinking about light as a wave helps us understand three properties of light.

Light can reflect. Reflected light lets us see all the objects around us. Reflection also allows us to see colors. When light hits an object, some of it is absorbed. Some of it bounces off, or reflects. When light hits a smooth object, it bounces off it at the same angle but in a different direction.

Light can scatter. When light hits a rough object, it bounces off it in all different directions. Paper is not perfectly smooth like a mirror. The way light scatters when it hits the paper allows you read the words on it.

Light can refract. This happens when light passes from one object to another object. Light hits the second object and changes both direction and speed. A straw in a clear glass of water looks broken because of refraction. Light hitting the part of the straw above the water passes only through air. Light hitting the underwater part of the straw goes from air to water. The light reflects off that part differently than off the above-water part.

Refraction is the reason that these straws look bent or broken.

Early scientists began to map what light waves look like. Wavelengths measure the size of light waves. A wavelength is the distance between one crest (the highest point of a wave) and

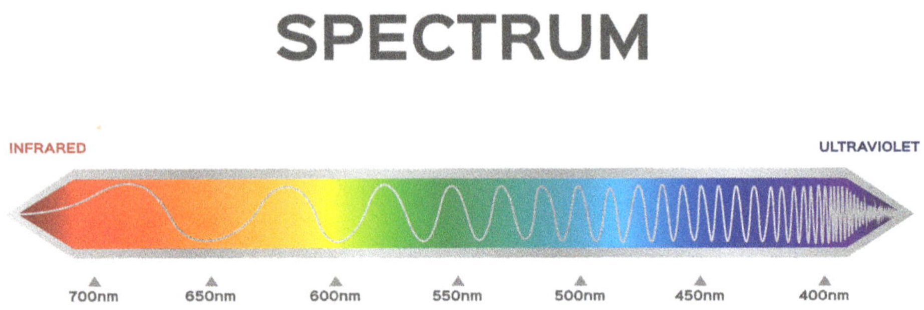

Different wavelengths and frequencies create the visible light spectrum. This is why we see different colors.

the next crest. Light's frequency is the number of crests that pass a point in one second. Frequency's unit of measurement is the hertz. Color results from the frequency of the type of light that we can see. Red is at one end of the color spectrum. Its light waves move at a frequency of 430 trillion hertz. When you see violet, you are seeing light waves that move at 750 trillion hertz.

Light as Particle

In 1905, Einstein noticed that light was full of photons. Photons are particles of energy. Light sources produce huge numbers of photons. Objects in a room reflect those photons. Therefore, photons help us see objects. Einstein was describing light as a particle. However, he did not ignore the wave definition of light. He suggested a wave-particle duality. This means something can behave both like a wave and like a particle. Scientists today agree with him. Light is particles traveling as waves.

MAKE YOUR OWN MODEL: OBJECTS REACTING TO LIGHT

A sunny day works great for this model. You'll need to be able to put items in a sunny yard or on a sunny sidewalk. It is also fun to do this activity with a sibling, parent, or friend. Scientists usually work in teams. They work with others because everyone notices different things, asks different questions, and suggests different ideas. Working together with someone else, you can learn a lot more than you would by yourself.

Materials

paper to write on

pen or pencil

shadow (If you do this on a sunny day, you should be able to find a shadow that is near your experiment.)

book

window to look through

mirror

magnifying glass

empty clear glass

wax paper

other small items found around your home or classroom that you can move

Directions

1. Together, write down definitions to these five words: *transparent*, *opaque*, *translucent*, *reflectivity*, and *refractivity*. Use a dictionary if you need to.
2. Examine each object—book, window, mirror, magnifying glass, empty drinking glass, and wax paper—as well as the shadow. On your own, answer these questions: Is each transparent, opaque, or translucent? Does it reflect light? Does it refract light?
3. Share your notes with your fellow experimenter. Do you agree on everything? Do you disagree on anything? Can you make a good case for why your answers are correct?
4. Use what you learned from this model to think of other objects and materials that might fit these classifications. See if you can find any other objects to compare to the original model.
5. Discuss what you thought of and what other objects you found.
6. Put the items in order from most transparent to most opaque.

7. Experiment more with reflection: Stand near the mirror but not directly in front of it. Notice how you can see each other in the mirror even though neither of you are standing right in front of it.

MAKE YOUR OWN MODEL: OBJECTS REACTING TO LIGHT

Now let's do some bonus experimenting. We experience light waves every time we see something. We also experience them during another common situation: when we touch something that has been sitting in the sun or under a lamp. We feel that the object is hot. Light is what causes that feeling of heat.

In sunlight, will a white T-shirt or a black T-shirt feel warmer?

Materials

several T-shirts:
at least one that is mostly black and one that is mostly white, and a few others that are each mostly other colors, like gray, red, light blue, and dark blue

a sunny spot

Directions

1. Take the T-shirts outside and spread them on the ground in the sunshine.
2. Take a short break to let the shirts sit in the sun.
3. When you come back outside, touch each T-shirt. Does one feel warmer than another?
3. Name other objects and state which ones you think would be warmer or cooler based on this model.

Test Yourself

1. What are easy-to-remember definitions for transparent, opaque, translucent, reflectivity, and refractivity?
2. How did each object interact with light?
3. How do mirrors work?
4. Why did dark-colored T-shirts feel hotter in the sun than light-colored ones?

Answers:

1. Transparent means clear; you can see through it easily. Opaque means you cannot see through it at all. Translucent is kind of in the middle of transparent and opaque—you can see through something that is translucent, but you cannot see through it clearly. Reflection means the light comes back toward you. Refraction means the light is bent inside the object.
2. Shadow: Opaque, may reflect
   Book: Opaque, reflects
   Window: Transparent, may reflect
   Mirror: Opaque, reflects
   Magnifying glass: Transparent, refracts
   Empty glass: Transparent, may reflect
   Glass with water and straw: Transparent (glass and water) and opaque (straw), refracts
   Wax paper: Translucent, reflects
3. A mirror reflects light back at the same angle that it enters the mirror.
4. Darker objects absorb more light than lighter objects, which is why they feel warmer in the sunlight. A black T-shirt absorbs all light waves, but that light does not just disappear in the shirt's wrinkles. It gets changed into heat energy. A white T-shirt absorbs almost zero light, so it has nothing to change into heat. It then does not feel as hot in the sun.

Sound waves are behind everything we hear, even whispered secrets!

CHAPTER 3

Modeling Properties of Sound Waves

Modeling can help you at all stages of an experiment. Modeling helps scientists ask questions, answer questions, and describe what they learn. Sometimes scientists model early, before they know much about a subject. You can use modeling to learn more about sound with a game of Simon Says. You may have played Simon Says before. During the game, one person gives instructions to their friends. They tell their friends how to move. Those friends must follow those instructions perfectly to win. However, only instructions that start with "Simon Says" should be followed. Otherwise, you're out! Because talking and listening are so important, this game is a great adventure in sound. Also, with a little adjustment, it can help you remember the definitions of two important parts of sound: wavelength and amplitude.

MODELING WAVELENGTH AND AMPLITUDE

For this activity, you do not need any materials, just yourself and at least one other person. The game will get you thinking about how sound travels. As you move forward and up and down, you will imagine how sound waves move like that too.

Pay close attention to instructions you hear. This will also help you think about how sound works. Talking and listening seem to take no work at all. Some people talk in their sleep! Sometimes you can listen even when you are doing something else. However, in reality, your body and sound waves are working hard together to translate sound into words.

How the Ear Hears

During Simon Says, your partner uses their vocal cords, tongue, lips, and jaw to form words. All this movement vibrates the air around their mouth. This vibration travels, as sound waves, through the air to your ear. The waves enter your outer ear and then travel along your ear canal. That is like a skinny hallway to

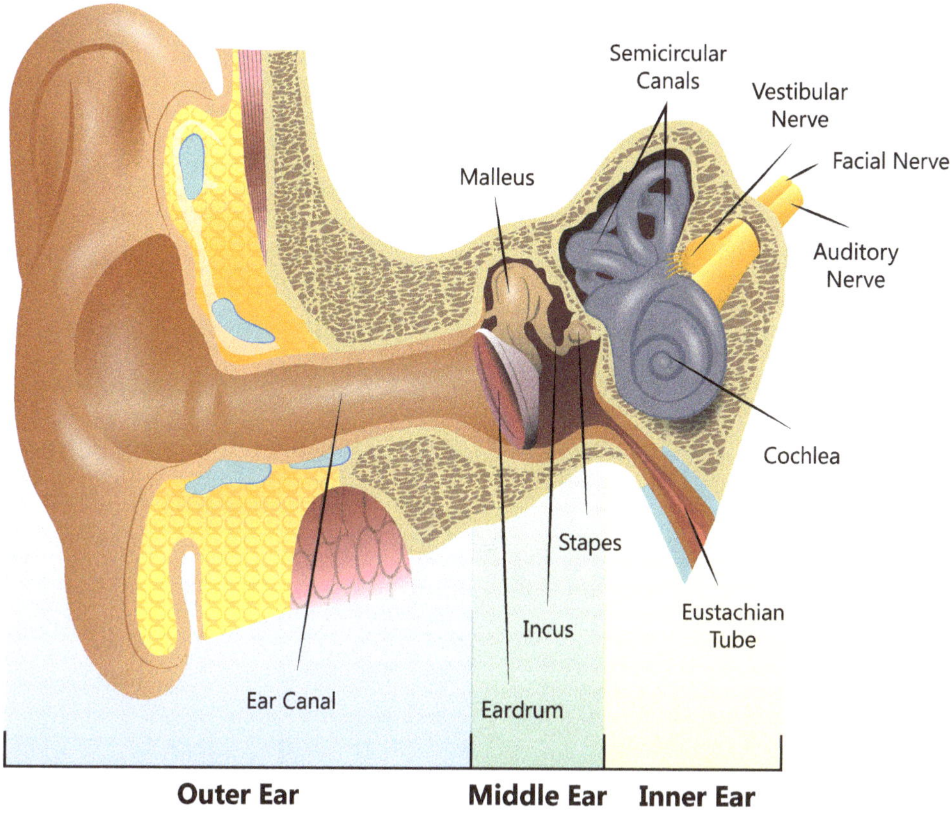

This diagram shows the important parts of your ears and ear drums.

your eardrum. The sound waves reach your eardrum. It vibrates. This then vibrates three tiny bones in the middle of your ear. They send the vibrations on to your inner ear. There is fluid in there. It ripples with the vibrations. Hair cells in the ear are next to catch the sound wave. Their movement triggers a rush of chemicals. These create an electrical signal. One of your nerves carries this signal to your brain. Finally, you hear, and understand, your partner's words—but you also need to pay close attention to what you are hearing!

What the Ear Hears

It is helpful for this model to know what sound waves look like. This model explores two parts of the wave, wavelength and amplitude. Wavelength is the distance between two crests, or high points of a wave. Wavelength is directly connected

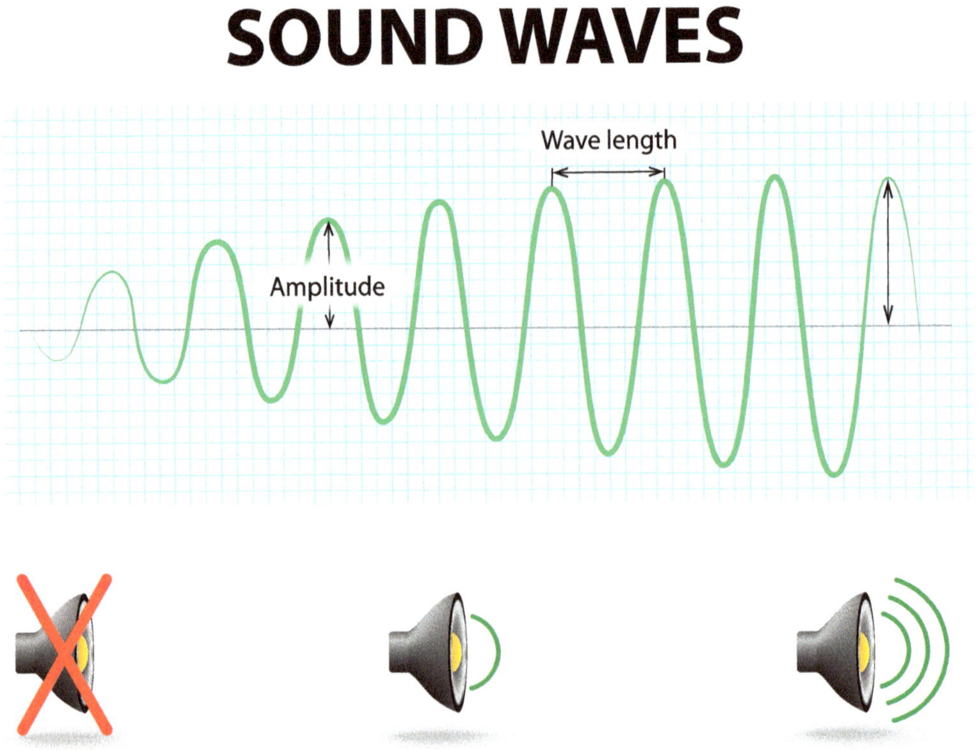

Longer wavelengths have a lower frequency. Larger amplitudes make for louder sounds.

to what kind of sound we hear. Longer wavelengths—longer distances between crests—have lower frequencies. They have lower pitches. A bear's growl and the rumble of a car have longer wavelengths. Shorter wavelengths—shorter distances between crests—have higher frequencies. A squeaky dog toy and the siren on an ambulance have shorter wavelengths.

Amplitude is the measurement of how much a wave changes from its place of rest. Amplitude is associated with how loud or soft a sound is. If a sound wave's amplitude is a high number, the sound is loud. Amplitude is measured in decibels.

When you turn up the volume on the television or your computer, you increase its decibels. Decibels measure sound's loudness and softness. A whisper measures 15 or 20 decibels. When your friend played Simon Says with you, their instructions were probably at about 65 decibels. Concerts can be as loud as 120 decibels. Fireworks are about 150 decibels.

Your ears will start to hurt when the sound reaches 130 decibels. Loud sounds can actually injure your ears. They hurt the hair cells that release chemicals in your ears. If those cells get really hurt, they may die. They will not regrow. This can lead to loss of hearing. Even sounds as quiet as 85 decibels can do damage. (Traffic during rush hour can reach 85 decibels.) The longer and more often you listen to loud noises, and the

Really loud sounds, like fireworks, can damage your ears. That's why ear plugs are a good idea at concerts or fireworks displays.

closer you are to them, the more likely you are to hurt your ears. Don't turn your headphones up too loud. If you are going to be around loud noises for a long time, it is a good idea to wear earplugs.

Directions

If you have played Simon Says before, you know there is a lot of movement in the game. There is movement forward. For example, the person who is giving instructions says, "Simon says, 'Run ten steps.'" There is movement up and down. For example, "Simon says, 'Hop until I tell you to stop.'" Sound moves forward and up and down too. In this game of Simon Says, you will use sound words for moving. Instead of talking about running or walking, you will use the word "wavelength." Instead of talking about jumping or hopping, you will use the word "amplitude." In this game, you will talk about wavelengths for moving forward and amplitude for moving up and down.

Here are some examples of what you and your partner can do:

- Your partner says, "Simon says, 'Move with a really big wavelength.'"

  You take big steps forward.

- Your partner says, "Simon says, 'Move with a small amplitude.'"

  You take tiny hops.

- Your partner says, "Simon says, 'Walk with a small wavelength and a big amplitude.'"

  You take little steps. At the same time, you should stretch your neck up high and then bow your head down low, over and over. Or squat and pop back up as you take small steps forward.

- Your partner says, "Simon says, 'Have no amplitude.'"

  You lie down.

The Speed of Sound

When you played this game, was the day hot? Was it cold? Were you and your friend in the same room? Was there a door between you? Where you were had an effect on how fast and far the sound of your friend's voice traveled.

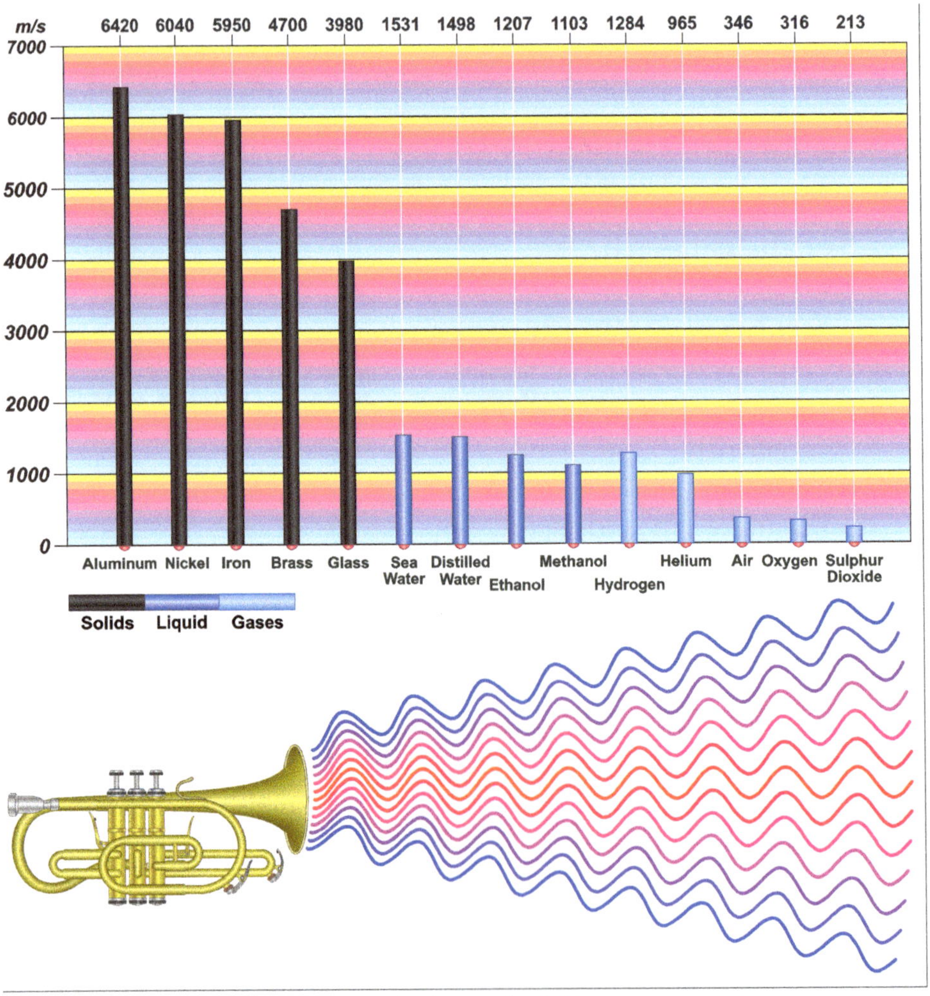

Sound travels at different speeds through different materials.

Sound waves travel at about 768 miles per hour (1,236 kmh) through the air. They move faster through warm air than cold air. They move faster still through liquids and solids. That is because the molecules in the air on a hot day are moving fast. (Air molecules are the bits of gases that make up our air.) Therefore, these fast-moving molecules can move the sound waves along fast. Sound moves fastest in solids. A solid object is dense. Its molecules—the small components that make it what it is—are close together. When the sound wave vibrates the object, it takes less time for each molecule to bounce against its neighbor. It takes less time for the sound wave to travel through the solid.

When there is cold air, sound can travel farther. Often there is a layer of warm air above the cold air. Sound waves bend, or refract, when they hit the different temperatures. This can send the sound downward. It may reach someone standing on the ground far away.

The speed of sound is much slower than the speed of light, but still, very few things can move faster than the speed of sound.

Most planes do not even fly faster than the speed of sound. Fighter jets can. This is called flying above Mach 1. (Mach 1 is the speed of sound. Mach 2 is twice the speed of sound.) It is

When a plane is about to break the sound barrier, changing air pressure around the plane creates a vapor cloud.

also called breaking the sound barrier. A plane that flies this fast creates a sonic boom. Sound waves being forced together so fast and so powerfully make this noise. It can sound like an explosion!

Test Yourself

1. What are you measuring when you measure the distance between crests?
2. What wave measurement tells you the strength of a wave (how loud or how soft a sound is)?
3. If you want your friend to hear your new favorite song—right now!—should you play it for them underwater or while sitting on the beach?
4. Can sound physically hurt you?

Answers:

1. wavelength
2. amplitude
3. The sound waves will travel faster to your friend underwater than through the air. However, we hear sound waves moving through the air better than those moving through the water.
4. Yes! Loud sounds that are nearby can damage the inside parts of your ears. That can damage your ability to hear.

Lasers are silent. Military lasers have added sound effects so that operators can tell when lasers are firing.

CHAPTER 4

Seeing and Feeling Sound

Because we understand how light waves and sound waves act and react, we can use them to help improve our lives. Scientists are using light and sound to reach many unique goals. Some are big goals, like predicting natural disasters. Some are even bigger than that, like finding alternatives to electricity.

TSUNAMIS AND SOUND WAVES

Researchers have detected tsunamis with expensive buoys placed all around the world. Tsunamis are long, tall ocean waves. They are so powerful that they cause great damage when they hit land. If people living along the coast are given a warning

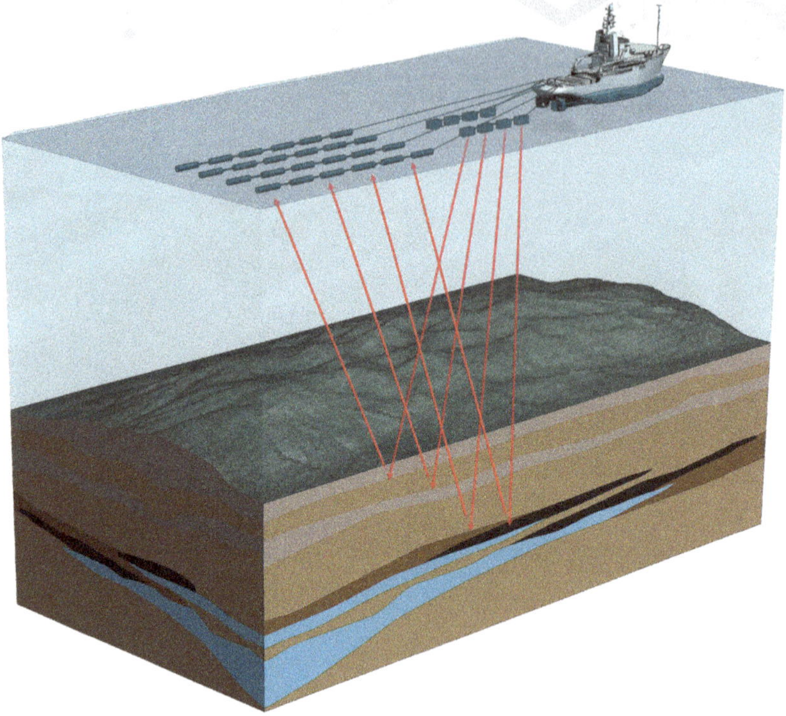

Hydrophones help people hear sound waves traveling underwater. These sound waves help us map the ground far below the water's surface.

that a tsunami is coming, they can get out of the way. Lives and property can be saved. However, waiting for the buoys to measure pressure changes in the ocean caused by tsunamis takes too long. By the time the wave reaches the buoy, it is too late to protect people.

Now, scientists and mathematicians think they may be able to use sound waves to predict disasters along Earth's coasts. Earthquakes that happen under the ocean can cause tsunamis. Researchers realized that if they understood each earthquake, they could predict if it could cause a tsunami. They placed one hydrophone in the ocean. A hydrophone detects sound waves traveling underwater. Sound waves can travel more than ten times faster than tsunamis. They also travel in all directions from an earthquake. That means researchers need to set out only one hydrophone. One of the sets of sound waves will hit it, no matter where it is placed. The characteristics of the sound waves tell researchers about the earthquake: where it happened, how long it lasted, and how big it was. They can use that information to predict where and when a tsunami will hit. Minutes after the hydrophone hears the sound waves, a tsunami alarm can be sounded.

MAKING LIGHT WITH BUBBLES

While some scientists are studying big waves of water, others are studying tiny droplets. Even those small droplets could change our lives. We know that sound waves are powerful. They can damage ears. That means sound waves certainly can burst bubbles. However, bubbles are not just delicate spheres of gas. They, too, have power. When sound crushes them, they release energy. This creates a tiny, short burst of light and heat. Scientists have been studying this cycle when it is produced by a small, ocean-dwelling creature: the snapping shrimp.

Snapping shrimp use sound to stun their prey.

When the snapping shrimp closes its claw, the change in pressure forms bubbles. At the same time, the claw's snapping sound pops the bubbles. Under the popping pressure, the gas inside the bubbles heats. There is a reaction within the heating that creates light. This all takes less than a billionth of a second to happen.

Scientists have been using lasers—another form of light—to shoot bubbles. They hope to learn how to develop a new light source with bubble fusion. The current ways we make light also make a lot of pollution. Scientists are using their understanding of light waves to look for more sustainable and environmentally friendly ways of making light.

MAKE YOUR OWN MODEL: MOVE OBJECTS WITH SOUND

Now you will model something you may have your own practical experience with. Have you ever played your music, television, or video game so loudly that the picture frames on the wall rattled? In this model, you'll learn something about how sound can move objects.

Materials

mixing bowl

plastic wrap

uncooked rice (a small amount, about a spoonful)

cookie sheet (or other metal pan)

metal spoon

Any size bowl will work for this model.

Directions

1. Cover the bowl with a piece of plastic wrap. Make sure to pull the plastic tight over the bowl and that its edges stick securely to the bowl.
2. Pour the rice onto the plastic wrap.
3. Hold the metal pan in the air close to the bowl.
4. Drum on the pan with the metal spoon.
5. Your model is a success when the rice begins "dancing" to the beat.

FEELING SOUND

This model helps us to understand how sound is very much a part of everyone's lives—even the lives of people who cannot hear. Everyone is affected by sound waves.

When Heather Whitestone was just twenty-one years old, she won the Miss America pageant. It was 1994, and she became the first person who was deaf to win the competition. She had lost her hearing when she was eighteen months old. Whitestone surprised the pageant audience by dancing ballet to Sandy Patti's "Via Dolorosa" during the talent portion of the contest. People were amazed that someone who could not hear could move in time with the music.

In 2001, seven years after Whitestone's win, scientists explained how Whitestone's brain helped her "hear" the music. Dr. Dean Shibata led the research on this at the University of Rochester School of Medicine in New York. Shibata and his team used technology to compare the brain activity of people who can hear with the brain activity of people who cannot hear. They learned that people who are deaf sense vibration—which is caused by sound waves—in the part of the brain that other people use for hearing sound waves.

Students at the National Technical Institute for the Deaf, a college for people with hearing impairments, are also familiar with how sound-wave vibrations affect them. These students receive inflated balloons when they go to school concerts. They get inflated balloons to hold in order to feel the musical vibrations, which allows them to "hear" the music.

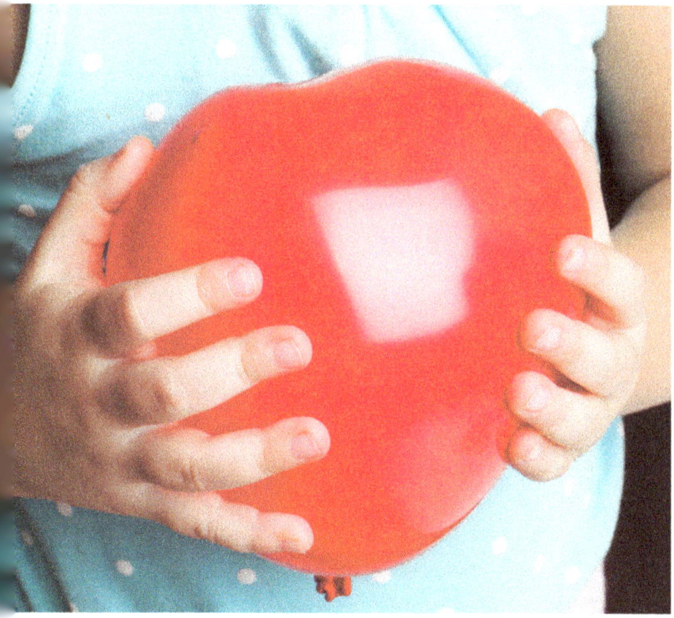

You can feel the vibrations from music by holding an inflated balloon at a concert.

Test Yourself

1. Why did the rice move even though you did not touch it?
2. Would the rice have moved if the grains had been placed directly in the bowl, not on plastic wrap?
3. Aside from moving the rice, what else did the sound waves from the spoon hitting the pan do?
4. How could someone who cannot see or hear tell sound was being created?

Answers:

1. The sound waves caused by you hitting the pan with the spoon traveled through the air. They made the plastic wrap vibrate. That caused the rice grains to move.
2. They could have—but only with a bigger sound than you could create with just a spoon and a pan!
3. They allowed you to hear the sound of the spoon striking the pan.
4. They could hold an inflated balloon near the spoon-and-pan drum. The sound waves from the spoon hitting the pan make the plastic wrap vibrate. They also would make the balloon vibrate. The person holding the balloon would feel those vibrations.

Glossary

amplitude The height of a wave.

crests The highest points of a wave.

frequency The number of crests in a second.

light Energy that allows us to see and that behaves like both a particle and a wave.

particle A very small unit of matter.

photon A bundle of energy.

reflection The act of waves bouncing back. When a sound or light wave hits an object, some or all of it may bounce off the object. It reflects off at the same angle as it hit the object. Reflection allows us to see an object.

refraction The act of waves bending. Sound and light waves refract when they hit an object and keep moving through it. For example, sunlight travels through air toward a lake. When it strikes the water, it continues to move through it, but it changes direction. It refracts when it hits the water.

sound A vibration you can hear.

wave A disturbance that travels through space and matter, carrying energy with it.

Further Information

BOOKS

Kenney, Karen Latchana. *Sound and Light Waves: Investigations.* Minneapolis, MN: Lerner Publications, 2017.

Rowell, Rebecca, and Alex Sciuto. *Energy and Waves Through Infographics*. Super Science Infographics. Minneapolis, MN: Lerner Publications, 2014.

Winterberg, Jenna. *Sound Waves and Communication.* Huntington Beach, CA: Teacher Created Materials, 2016.

WEBSITES

DK Find Out! Sound

https://www.dkfindout.com/us/science/sound

Learn more about sound, from making music to breaking the sound barrier, on this fun website.

Light and Color

https://digital.scetv.org/knowitall/nasa/light/light_color.html

On this website produced by NASA, interact with colors to learn how the eye sees each one.

VIDEOS

Two-in-One Photography: Light as Wave and Particle!

https://www.youtube.com/watch?v=mlaVHxUSiNk&feature=youtu.be

In 2015, scientists finally were able to photograph light appearing as both a wave and a particle. Watch a fun explanation in this video.

What Is Sound?

https://www.youtube.com/watch?v=3-xKZKxXuu0

This video uses animation and modeling to explain what sound is and how it works.

Index

About the Author

Kristin Thiel has written many titles for Cavendish Square Publishing, from books on science and technology, to biographies, to titles on history and current affairs. Her first was a biography of Dorothy Hodgkin, a Nobel Prize–winning crystallographer. She is the author of *Science Investigators: Investigating Forces and Motion Through Modeling.*